The Gravity Cycle

A Philosophical Hypothesis
of the Big Picture That
Eluded Einstein

I0476024

Martin O. Cook

Edited by Wendy A. Cook

Some Parts From Previous Versions Edited by Stephen Gibbons

Insightful Contributions by David Miller, Monte Mower, Jeffery Davis, **Wendy Cook**, Stephen Gibbons, Jack Hylton Jr.,Bryan McPherson, Randy Dean, Mike Scott, John Cook Sr., Sally Cook, Kylene Schramm, Thomas Cook, Sarah Cook, Rebekah Cook, Isaiah Cook, Kathryn Cook, and Kelsey Dean Grace Cook

"For decades Einstein attempted to develop a unified field theory... connecting the movement of planets and stars with the operations of the tiniest subatomic particles." (Lacayo, 2014, 9)

Hawking was involved in the search for the great goal in physics—a "unified theory." Such a theory would resolve the contradictions between Einstein's theory of relativity, which describes the laws of gravity that govern the motion of large objects like planets, and the quantum mechanics theory, which deals with the world of subatomic particles. For Hawking, the search was almost a religious quest—he said finding a "theory of everything" would allow mankind to "know the mind of God." ...In later years, though, he suggested a unified theory might not exist. (Bar, 2018.)

Introduction

How does one push through the minutia of tradition? For about a century, modern physics has been plagued by the chasm existing between general relativity and quantum mechanics. Some seek to bridge this gap with a mathematical solution in the form of a unifying equation, a quest initiated by Einstein. To this day, this gap-bridging equation remains hidden, or, as Stephen Hawking suggested, might not even exist at all. What if the bridge that could unify the science of gravity with the science of atomic and quantum processes requires a change to the very foundation of physics established by Galileo and Newton?

The difficult part of introducing a different way of thinking that could unify the oil and water mix of space-time gravity with our advancing understanding of atomic and quantum processes is the countless potential contradictions waiting in the shadows and cracks of a developing perspective. Unification requires changes. Changes do not come easily. By exploring the possibility of a major paradigm shift in the very foundation upon which we presently view the momentum, relativity, and gravity of masses, we may eliminate the chasm that left Einstein stumped and continues to leave present theoretical physicists with the same results encountered by their mentor.

To this day, modern physics lacks an atomic/quantum model of motion to explain the role of atoms in the momentum, relativity, and gravity of masses. External forces can speed masses up or slow them down, but there is no explanation to account for what is happening on the inside, on the atomic level. This lack of accountability and equilibrium of energy in the changes of motion of any mass is referred to in this book as the motion myth paradigm and is akin to the pre-Lavoisier days in the science of chemistry. An atomic model of motion is needed to dispel the motion myth paradigm in order to unite the interactive nature of all the known forces.

The four parts and one appendix of this book are as follows:

Part I: The Gravity Cycle

The first part introduces the gravity cycle. Simple illustrations show gravity as a cyclical flow of energy circulating within a galaxy and throughout the universe.

Part II: The Motion Myth Paradigm

The second part introduces the motion myth paradigm as an outdated perspective hindering the development of an atomic model of motion.

Part III: The Atomic Model of Motion (AMM)—The Role of the Atom in Momentum, Relativity, and Gravity

The third part of this book introduces an atomic model of motion. It encompasses what momentum, relativity, and gravity look like when atomic and quantum processes are integrated into our understanding of them.

Part VI: The Galaxy Gravity Cycle (GGC)—The Role of Spherical Bodies in Absorbing and Redirecting the Flow of Gravitational Energy within a Galaxy

The last part of this book explores spherical bodies as the means of absorbing and redirecting the flow of gravitational energy within a galaxy and throughout the universe.

Appendix A: The Problem with Special Relativity

The appendix addresses the problem with special relativity when viewed through the lens of an atomic model of motion perspective.

Before you begin to read about the gravity cycle, I think it is very important to discuss my use of the term *gravitational energy*. In the past, I have talked about gravitational energy as an undefined and undetected form of energy that plays a vital role in the momentum, relativity, and gravity of atoms and masses. Without empirical proof or at least some hint of the existence of gravitational energy, the gravity cycle as a hypothesis has little chance of penetrating and redirecting the current scientific understanding of gravity as defined in general relativity. Recently, someone asked me how my hypothesis of the gravity cycle takes into consideration what scientists are now calling dark matter. Honestly, I really didn't know a lot about dark matter. Around this same time, I came across a passage in a book that my wife had recently given to me.

> In addition, a striking limitation of the Standard Model has appeared in recent years. Around every galaxy astronomers observe a large cloud of material which reveals its existence via the gravitational

pull that it exerts upon stars, and by the way it deflects light. But this great cloud, of which we observe the gravitational effects, cannot be seen directly and we do not know what it is made of. Numerous hypotheses have been proposed, none of which seem to work. It's clear that there is *something* there, but we don't know what. Nowadays it is called 'dark matter'. Evidence indicates that it is something *not* described by the Standard Model, otherwise we would see it. Something other than atoms, neutrinos or photons. (Rovelli, 2014)

After reading this passage, it occurred to me that my hypothesis about the gravity cycle already addresses what could be the mystery of dark matter. Dark matter might be nothing more than my use of the term *gravitational energy*—a *dark* energy invisible to our means of detection but observable in the altered movement of light and masses. What if the effects contributed to dark matter is nothing more then the continuous flow of *gravitational energy* between galaxies as described in step 5 of the gravity cycle?

Gravitational energy, as presented in this book, is an allusive form of energy that cannot be directly encountered with our present instruments of detection, and yet, it is the underlying theme that ties all the sections of this book together. In part one of this book, gravitational energy is introduced as the *dark* energy that drives the galaxy cycle. Part two of this book points out that our lack of knowledge of the existence of gravitational energy and its role in the momentum, relativity, and gravity of masses sustains the mindset of a motion myth paradigm. Part three of this book describes gravitational energy as a unifying energy that drives the momentum, relativity, and gravity of masses. And as hypothesized in part four of this book, spherical bodies such as stars, planets, and moons are the means of redirecting this energy in all directions within a solar system, throughout a galaxy, and across the universe.

As we wrestle with the unification of general relativity and quantum mechanics, I predict that what we now call dark matter will eventually be labeled as a *dark* energy that freely flows between galaxies, causing the gravitational shifts described in the previous quote. This *dark* energy, referred to in this book as gravitational energy, is the key to understanding the role of atoms in the momentum, relativity, and gravity of masses. It is the key to unlocking gravitational acceleration from an atomic/quantum perspective. And finally, it is the key to uncovering an alternative perspective with the same effect as to what Einstein referred to as warped space.

Part I

The Gravity Cycle

The Gravity Cycle is a repeating pattern that keeps the flow of gravitational energy moving throughout a galaxy. Just as evaporation and precipitation are part of a perpetual process we call the water cycle, momentum and acceleration are part of a cyclical process that can be referred to as a gravity cycle. As with other cycles, the gravity cycle can be broken down into steps to simplify and explain how it works. By breaking the gravity cycle into steps, we can see the big picture of gravity and how it maintains the acceleration of masses that governs life on planets, the movements of solar systems, and the structures of galaxies.

Step 1: Gravitational energy continuously flows out from the center of a galaxy. Stars already in motion absorb gravitational energy. As gravitational energy is continuously absorbed into the atoms of a moving star, the star experiences a continual acceleration shift towards the direction of absorption. A star's absorption of gravitational energy is the gravity that keeps its momentum in an orbital pattern around the center of a galaxy.

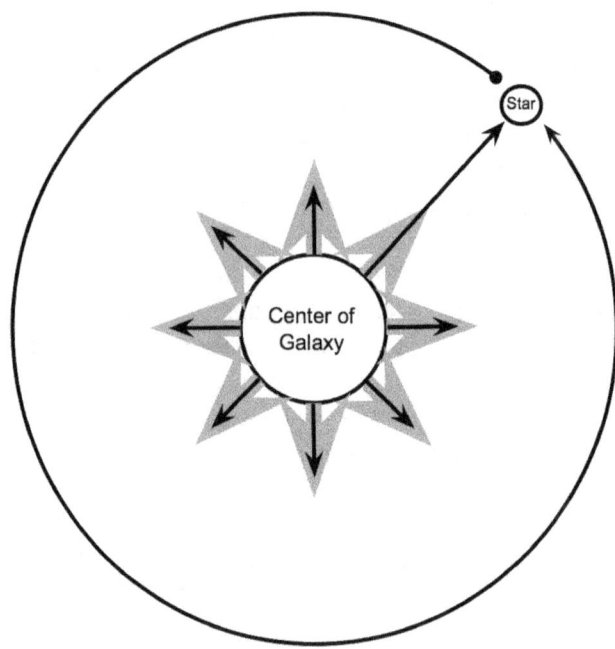

Step 2: As a star continuously absorbs gravitational energy, it also continuously emits it. It emits it proportionately through its spherical surface. The process goes like this: As gravitational energy is absorbed into the atoms of a spherical body, the atoms accelerate. Because space for acceleration within a spherical body is limited, gravitational energy is then released from these atoms in direct proportion to the amount they decelerate. Surrounding atoms absorb this gravitational energy, and again, due to limited space for acceleration, they decelerate and gravitational energy is released. This process of atoms accelerating and decelerating continues until the excess gravitational energy is eventually emitted proportionately through the spherical surface into open space. As this released gravitational energy travels through open space, it is absorbed into planets and other objects already in motion, accelerating their motion into an orbital pattern around the star's spherical shaped body, as the star's spherical shaped body accelerates in an orbital pattern around the center of the galaxy.

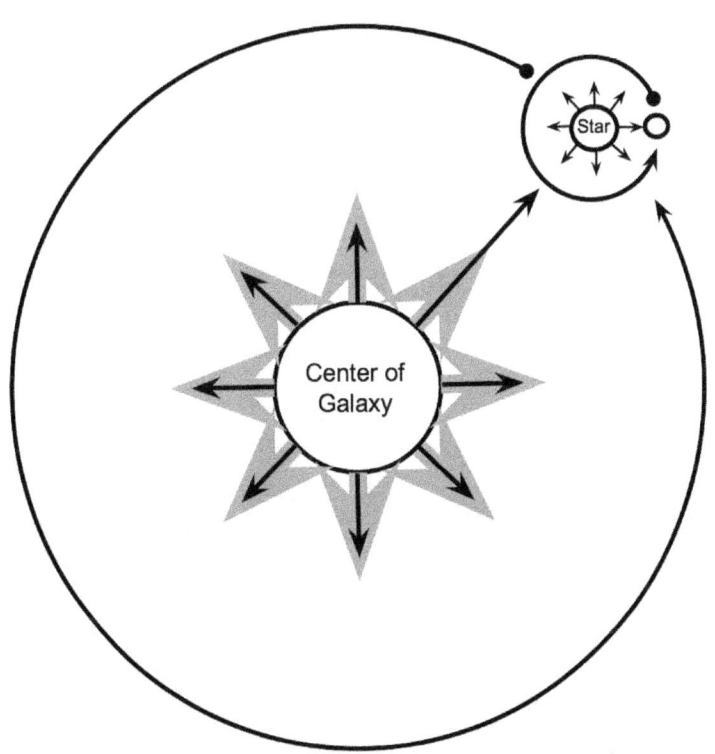

Step 3: Planets that are already in motion continuously absorb gravitational energy coming from the star they orbit. This gravitational energy keeps them accelerating in an orbital pattern around the star. As the Earth absorbs gravitational energy, it also emits a steady flow of gravitational energy through its spherical surface. The flow of gravitational energy emitted from its spherical surface causes other objects like the moon and man-made satellites to accelerate into an orbital pattern around it. The emitted gravitational energy also keeps objects like balls, cars, bikes, humans, and animals that are already in motion from floating off into space as the absorption of emitted gravitational energy causes them to continually accelerate into the planet's surface.

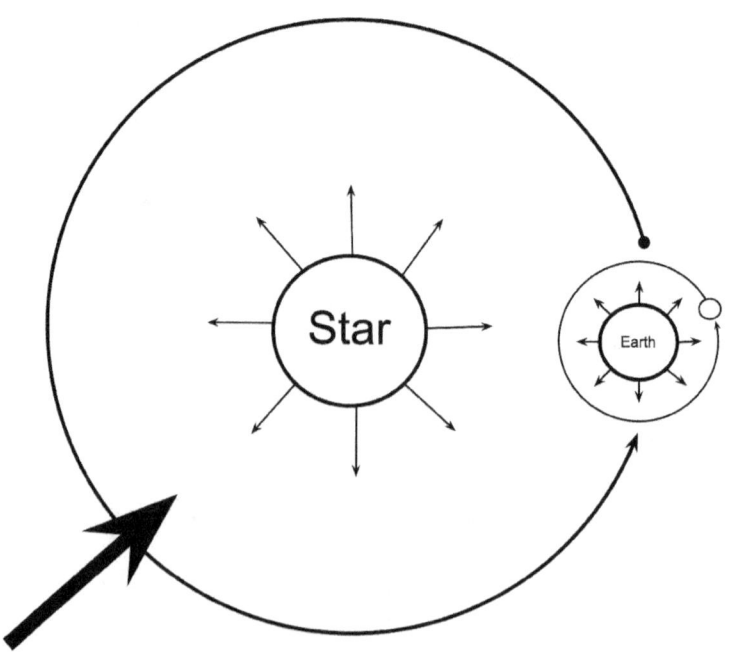

Step 4: Much of the gravitational energy that is emitted by the center of the galaxy makes it back to the center of the galaxy, where it is absorbed and redirected back out into the galaxy.

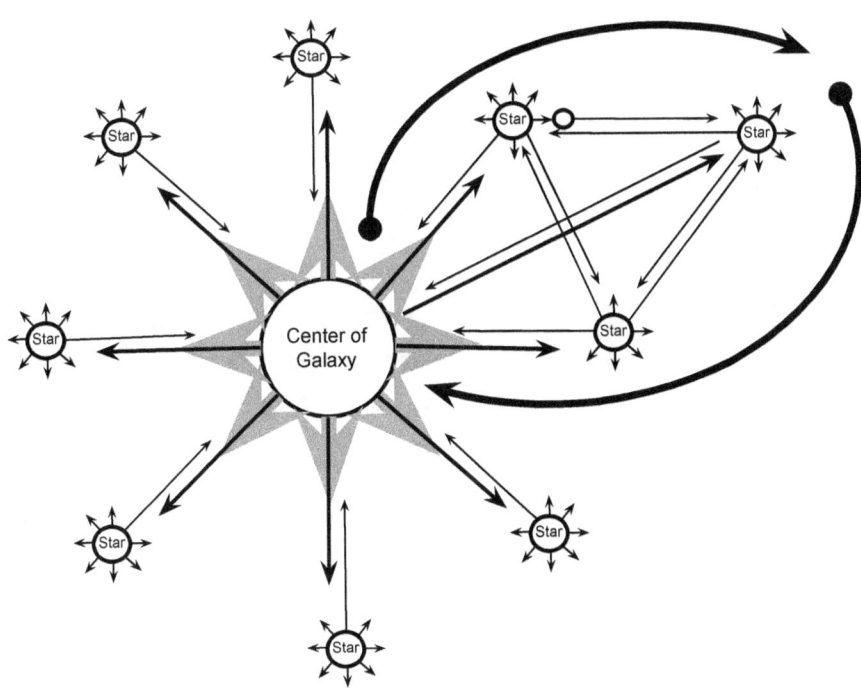

A complex web of spherical bodies absorbing and emitting gravitational energy sustains the gravity cycle within each galaxy.

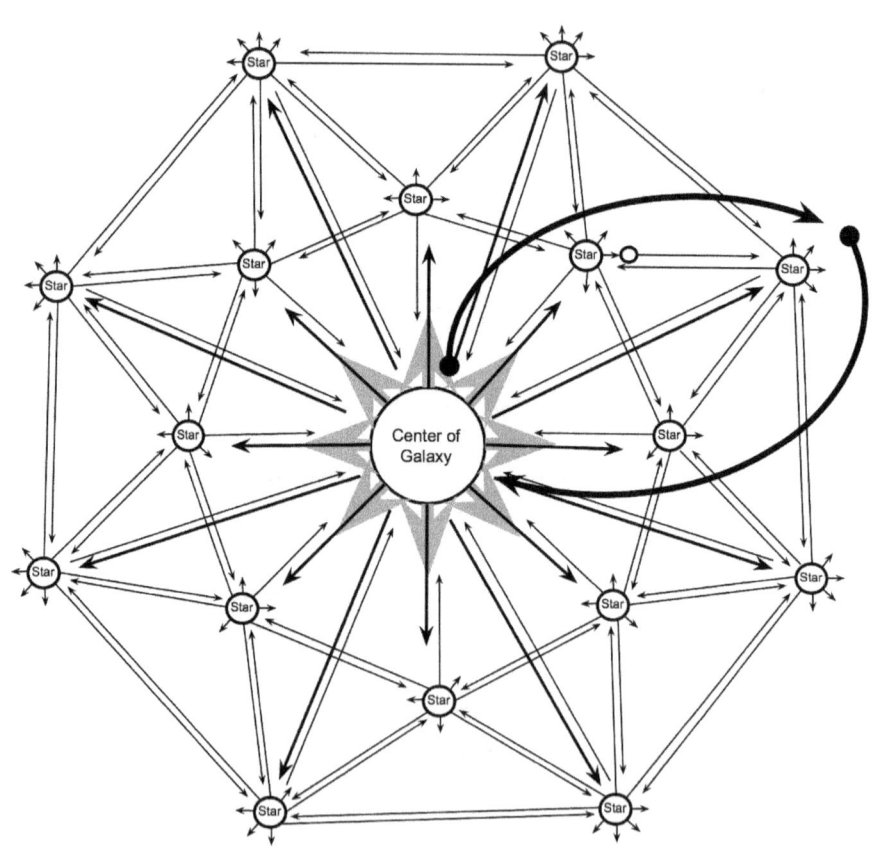

Step 5: Gravitational energy that escapes the gravity cycle of a galaxy feeds into other galaxies. In turn, galaxies receive gravitational energy from other galaxies to replace the gravitational energy they have lost. Our galaxy receives lost gravitational energy from other galaxies to replace its lost gravitational energy. Galaxies, like stars, play an important role in keeping the balance of gravitational energy flowing throughout the universe.

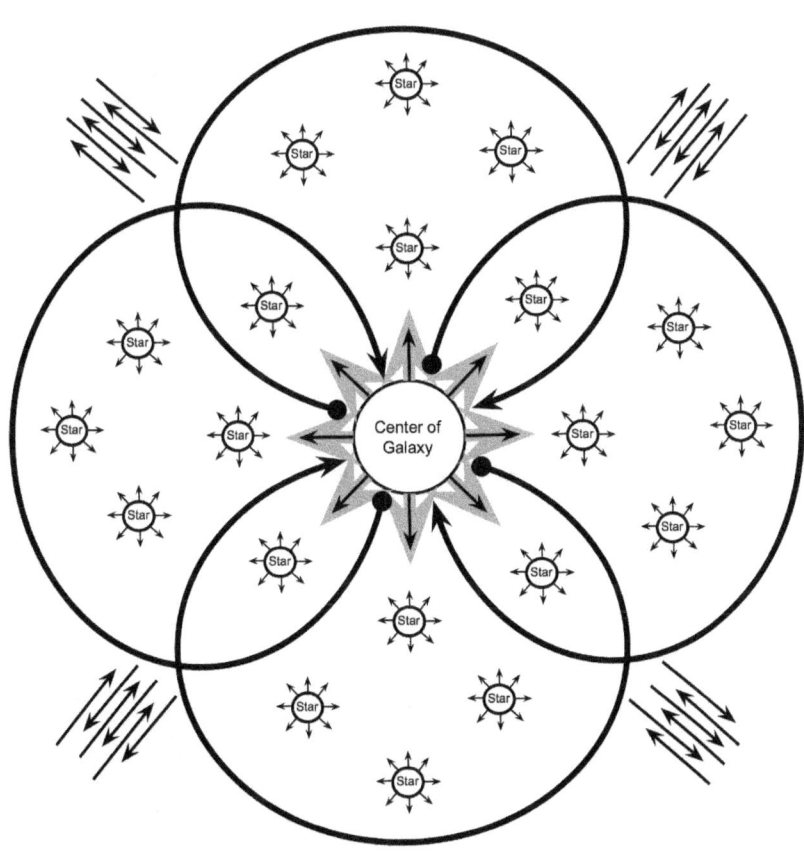

In summation, the cyclical flow of gravitational energy within a galaxy is the gravity that holds that galaxy together. Spherical bodies are the vehicles through which gravitational energy is absorbed and then redirected in all directions. As with any cycle, the process that keeps gravitational energy flowing within a galaxy repeats itself over and over and over again. And as some gravitational energy will escape the gravity cycle of our galaxy, our galaxy absorbs gravitational energy that has escaped the gravity cycle of other galaxies. The Gravity Cycle is the cyclical process that holds all moving bodies in orbit and keeps humans from floating off into space.

Part II

The Motion Myth

The motion myth is the false assumption that objects move freely through space without any atomic/quantum accountability. The motion myth stifles any attempts to unify the "outward" forces acting on masses to "inward" atomic/quantum processes operating within masses. The motion myth has its roots in Galilean relativity. It was then reinforced by Newton's first law of motion—the law of inertia, and then solidified by Einstein's relativities. As a result, when atomic/quantum physics started to mature in the early part of the twentieth century, it lacked a viable *model of motion theory* upon which to build and expand.

Galilean Relativity—Are there Free Rides?

Galileo reasoned that if he were confined in a windowless cabin of a uniformly moving ship, he wouldn't be able to tell the ship's speed relative to the land by applying what he knew about the laws of physics. In any inertial frame, if you throw a ball straight up, it will come straight down. Throw the same ball straight up in a different inertial frame, and you get the same result; it comes straight down. The same ball behaves the same way in different inertial frames. What Galileo saw with his eyes was correct—but what he didn't see tells the rest of the story.

Galileo failed to conceptualize that objects within an inertial frame are not getting a free ride. Free ride is the idea that objects appearing to be at rest within an inertial frame are piggybacking on the motion of that inertial frame. It is assumed that the resting object's motion is dependent on the motion of the object on which it appear to be resting rather than the resting object's motion being independent of the motion of the object or inertial frame on which it appears to be resting. The atomic/quantum processes of the atoms from which the resting object's mass is comprised drives its motion through space independent of the motion of the inertial frame wherein it appears to be resting. Because Galileo was completely naive to atomic/quantum processes operating within masses, he did not question the "invisible" workings of matter to account for the "visible" results observed. As the same ball goes from one inertial frame to another, atomic/quantum adjustments transpire within the ball to allow the relativity phenomenon observed by Galileo to occur. The ball is not getting a

15

free ride in either inertial frame. Its "invisible" inner workings change from inertial frame to inertial frame in order to yield the "visible" results observed in each differing inertial frame. Isaac Newton also failed to consider the "invisible" inner workings of matter when observing objects at rest while formulating his first law of motion.

Newton's Objects at Rest—Perpetuating the Motion Myth

When formulating his first law of motion, Newton differentiated between an object in motion and an object at rest. He observed that objects in motion tend to stay in motion and objects at rest tend to stay at rest. Newton didn't conceptualize that all objects are always in motion, even when they appear to be at rest. The subtle shift between recognizing and not recognizing the independent motion of masses, even ones that appear to be at rest, is the difference between developing or not developing an atomic/quantum theory that articulates the momentum, relativity, and gravity of masses. Newton viewed objects—whether in motion or appearing to be at rest—from their whole perspective and not from the "inner" parts or atoms that form each object. In doing so, he cemented the precedence that masses in motion tend to stay in motion and masses at rest tend to stay at rest with no thoughts as to how or why these phenomena maintain their consistency.

With the sole perspective that external, unbalanced forces speed masses up or slow them down, there was no investigation into the "inner" workings of masses that coincided with the external forces being applied to them. For Newton, this meant that masses could take a position of rest, a free ride perspective, similar to the mindset occupied by Galileo. This external mindset of motion prevented Newton from hypothesizing a theory to try to explain the "inner" science of motion to coincide with his "external" laws of motion. Such a theory would have attempted to explain how objects move through space in the first place—potentially exposing the motion myth, the idea that objects just mysteriously move through space without any "inner" atomic/quantum accountability—at the dawn of the scientific age.

Relativity Theories Perpetuate the Motion Myth Paradigm

"But when one starts with the wrong premise, no amount of patching can right the problem." (Seifer, 1996, 21)

16

This same misperception tainted Einstein when he brilliantly crafted his relativity theories. Einstein viewed masses from their whole perspective—just like Galileo and Newton before him. Einstein, later in his life, could not integrate atomic/quantum processes into his theories because he didn't attempt to balance "inner" processes with "outward" forces. Time dilation and space-time are byproducts of the motion myth paradigm and are mathematical substitutes for atomic/quantum processes. Einstein once said the following in referring to a mistake made by Max Planck:

> The main thing is the content, not the mathematics. With mathematics, one can prove anything. (Brian, 1996, 78)

Instead of exposing the motion myth, Einstein's relativity theories perpetuated it. This is where standing on Galileo and Newton's shoulders tainted his reasoning when it came to the motion of objects through space. He built off the faulty paradigm that masses just mysteriously move through space with no atomic/quantum accountability for changes in speed and direction. Einstein's relativity theories did not address the "inner" atomic/quantum processes driving the motion of masses, even masses that appear to be at rest. Alas, neither of Einstein's relativity theories exposed the motion myth paradigm, yet, both theories were brilliantly crafted within the limitations of this paradigm. [Special Relativity is addressed in Appendix A: The Problem with Special Relativity. General Relativity is addressed in Part IV under the subheading: How does emitted energy from spherical bodies better explain the warped space of general relativity?]

The combined works of Galileo, Newton, and Einstein collectively failed to address what should have been a basic fundamental question concerning the physics of motion: *How do masses move through space in the first place?*

The Motion Myth Paradigm

During a crucial part of a game, a coach will call a timeout to assess the situation and prepare his players to execute at the highest level, desiring the greatest outcome possible. Presently, in theoretical physics, a timeout is needed to assess the direction in which it is headed. Even with the apparent discovery of the Higgs boson, the Standard Model still falls short in explaining how momentum, relativity, and gravity operate from an atomic/quantum perspective. The failure to conceptualize that all objects are always in motion, even when they appear to be at rest, continues to erect an impenetrable barrier that prevents modern physicists from theorizing the atomic/quantum processes involved in the momentum, relativity, and gravity of atoms and masses. In order to rectify hundreds of years of tradition, we will carefully examine the motion myth and its profound impact on modern physics.

The following illustrates the consequence of the Motion Myth in perpetuating a paradigm that leads to a dead end.

The Motion Myth

Why did Einstein as well as the physicists standing on his shoulders fail to reconcile momentum, relativity, and gravity with atomic/quantum processes? "...when one starts with the wrong premise, no amount of patching can right the problem." Presently, modern physics is silent on the role of the atom in the momentum, relativity, and gravity of masses.

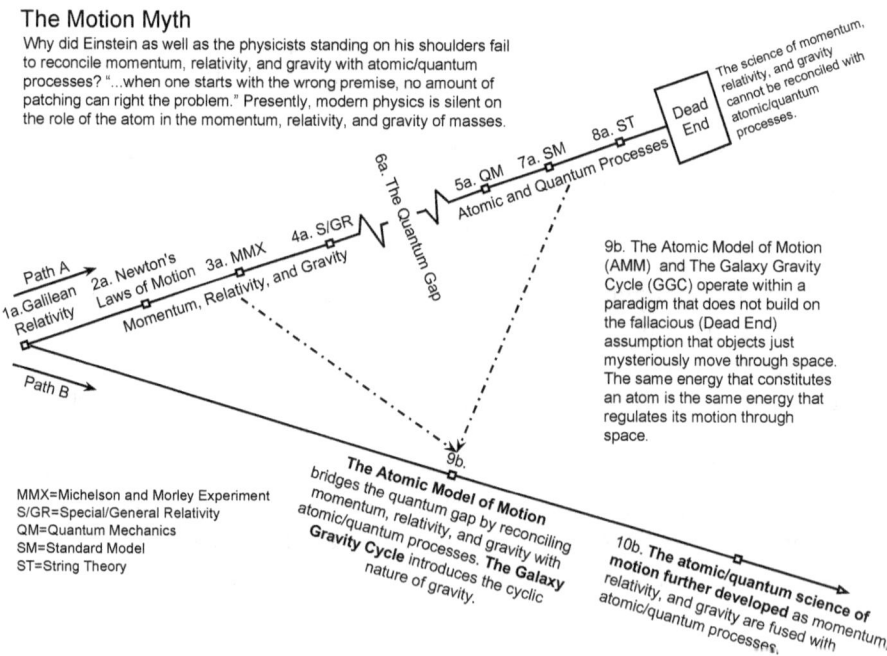

The science of momentum, relativity, and gravity cannot be reconciled with atomic/quantum processes.

9b. The Atomic Model of Motion (AMM) and The Galaxy Gravity Cycle (GGC) operate within a paradigm that does not build on the fallacious (Dead End) assumption that objects just mysteriously move through space. The same energy that constitutes an atom is the same energy that regulates its motion through space.

MMX=Michelson and Morley Experiment
S/GR=Special/General Relativity
QM=Quantum Mechanics
SM=Standard Model
ST=String Theory

9b. The Atomic Model of Motion bridges the quantum gap by reconciling momentum, relativity, and gravity with atomic/quantum processes. The Galaxy Gravity Cycle introduces the cyclic nature of gravity.

10b. The atomic/quantum science of motion further developed as momentum, relativity, and gravity are fused with atomic/quantum processes.

Path (a) represents the path we are currently traveling down. Any theory arising from the motion myth paradigm that attempts to connect "...the movement of planets and stars with the operations of the tiniest subatomic particles" will eventually run into a dead end.

Path (b) represents a new path that considers the role of atoms in the momentum, relativity, and gravity of masses. This path will lead to the further development of an atomic/quantum science of motion.

The following gives a brief description for the different points along the paths.

1a. (Galilean Relativity) The Motion Myth begins with Galileo's relativity observations when Galileo realizes that the same object in different inertial frames follows the same laws of physics. Galilean relativity goes forward without anyone seriously examining the "inner" workings of mass to coincide with the "outer" observations of this phenomenon—how the same ball behaves the same way in different inertial frames. When atomic/quantum physics comes onto the scene, it is not invoked as the key to reconcile the "inner" processes comprising matter with the "outer" observations of Galilean relativity. To this day, Galilean relativity cannot be explained from an atomic/quantum perspective.

2a. (Newton's Laws of Motion) Newton observed that objects appear to obtain a position of rest and included this observation in his first law of motion. The built-in assumption is that the object at rest is getting a "free ride" on the object it appears to be resting upon. Like Galileo, he makes no attempt to connect the "inner" processes comprising these objects with his "outer" observations. Newton's one-sided perspective—of viewing the motion or resting of objects from their whole perspective with no consideration of the operations of the individual parts from which these object are pieced together—sustained and perpetuated the motion myth that began with the one-sided view of Galilean relativity.

3a. (MMX) The Michelson and Morley Experiment in 1887 played a pivotal role in the direction of modern physics. Prior to this experiment, scientists hypothesized that an undetectable medium permeated space for the sake of facilitating the movement of light energy. The Michelson and Morley experiment was designed to detect the velocity of the Earth through this medium. The results of the experiment indicated that there was no medium. Both Michelson and Morley thought their experiment failed. A second attempt yielded the same outcome. For seventeen years, the scientific community grappled with this unexpected result. During this time, Fitzgerald and Lorentz composed separate contraction formulas to explain how a contracting apparatus in the direction of movement could yield the unexpected results.

Trying to account for Michelson's failure to find any movement of the earth in relation to the ether, Irish physicist George Francis Fitzgerald suggested that the measuring instruments Michelson used had contracted slightly and distorted the reading. He then produced equations showing that matter contracts in the direction of its motion, the contraction increasing as the speed increases. The Fitzgerald contraction, as this phenomenon is known, is "exceedingly small in all ordinary circumstances... If the speed is 19 miles a second — the speed of the earth around the sun — the contraction in length is 1 part in 200,000,000, or 2 ½ inches in diameter of the earth."

Dutch physicist H. A. Lorentz stated that a flying charged particle foreshortened in its direction of travel would increase in mass. Einstein, in turn, applied Lorentz's equations, known as the Lorentz transformation, to all objects, including clocks and measuring instruments. Einstein showed that objects moving at great speeds and over vast distances decreased in size and increased in mass. Strangest of all, he proved that at those speeds time slowed. (Brian, 1996).

In 1905, Einstein hypothesized that a permeating ether was no longer needed to explain how light moved through space. The majority of the scientific community adopted his perspective of the quanta (photon) nature of light. Of the two contraction formulas, Einstein adopted the Lorentz transformation to help mathematically orchestrate special relativity. From this point forward, time dilation put an end to the Michelson and Morley dilemma that had perplexed scientists for seventeen years. Time dilation became a permanent substitute for "inner processes" that may have caused the contraction of the apparatus in the direction of motion as hypothesized by both Fitzgerald and Lorentz. Time dilation sustained the continuation of the motion myth paradigm. In the end, the Michelson and Morley experiment did not end the rigid misperception that objects just mysteriously move through space.

4a. (S/GR) Special relativity and General relativity were both composed within the limitations of a motion myth paradigm. From that standpoint alone, both theories were destined to fall short in their ability to align with atomic/quantum processes.

In special relativity, Einstein applies the Galilean principle that the laws of physics are the same for all inertial frames to Maxwell's constant for the speed of light without understanding how Galilean relativity operates from an atomic/quantum perspective. The motion myth paradigm limited his perspective concerning the "inner" workings of matter in the motion of masses. This prevented him from considering the possibility that photons might not possess the same quality of atoms in adapting to differing inertial frames. In appendix A, I specifically address this problem by exploring the question, "Does a photon have the same relativistic qualities of an atom?"

In general relativity, Einstein deals with objects as a whole, using the enigma of space-time to initiate changes in motion, meaning gravity. His mathematical equations seem to validate the reality of this distortion. But when Einstein tries to connect "...the movement of planets and stars with the operations of the tiniest subatomic particles," he never succeeds. And no one after him has succeeded. The section in this book on "Quantum Gravity: The Role of the Atom in Gravity" explores how space-time translates into differing concentrations of flowing gravitational energy, and how this energy initiates changes in the motion of masses. The absorption of energy into every atom of a mass initiates and sustains the acceleration that we call gravity. The motion myth paradigm blocked the potential of seeing an alternative of space-time as differing concentrations of flowing energy accelerating the motion of already moving masses.

Like Galileo and Newton before him, Einstein viewed the movements of objects through space from a motion myth paradigm, stifling any attempt of "connecting the movement of planets and stars with the operations of the tiniest subatomic particles."

5a. (QM) The atomic/quantum revolution that began in the late nineteenth century and continued into the twentieth century was brought about by great minds such as Faraday, Maxwell, Michelson, Lorentz, Fitzgerald, Einstein, Rutherford, Bohr, Dirac, Heisenberg, Pauli, Schrodinger, and Feynman to name only a few. This is not an effort or inquiry pursued by a single person but rather the cumulative *contributions of many. In the pursuit to uncover mysteries of the universe through examining atomic/quantum processes, the role of the atom in the momentum, relativity, and gravity of masses didn't enter into the equation. This exclusion stems from a paradigm weaved into the foundations of modern physics—the motion myth paradigm. If only for this reason, quantum mechanics, the mathematical description of the motion and interaction of subatomic particles, will remain an incomplete science that will not be able to unite with the physics of motion until this issue is addressed and resolved.*

*6a. **The quantum gap** is the natural separation between the physics of motion including momentum, relativity, and gravity (the science of moving bodies) and the physics of atomic/quantum processes (the science of atoms and subatomic particles). Einstein's goal of attempting "to develop a unified field theory..." was his attempt to link or bridge the differences that create the quantum gap. Quantum mechanics has yet to be linked to Galileo's relativities, Newton's laws of motion, or Einstein's Relativities. The main culprit for this gap is the underlying impact of a motion myth paradigm mindset.*

7a. and 8a. (SM and ST) The Standard Model and String Theory are attempts to bridge the quantum gap but are also the byproducts of a motion myth paradigm. "But when one starts with the wrong premise, no amount of patching can right the problem." (Seifer, 1996, 21).

9b. The Atomic Model of Motion (AMM) and the Galaxy Gravity Cycle (GGC) combine to provide a simple foundational theory that ties together Galileo's relativity, Newton's laws of motion, Einstein relativities, and atomic/quantum processes. By exposing the stifling influence of the motion myth paradigm, a link can be made and equilibrium established and maintained between the "outer" phenomena of momentum, relativity, and gravity to the "inner" workings of atomic/quantum processes. This is the first step in the development of an atomic/quantum model of motion.

The Motion Myth paradigm remains the main obstacle preventing modern physicists from surmising how atomic/quantum processes drive the momentum, relativity, and gravity of masses.

Quantum Physics is Incomplete

"I like to think that the moon is there even if I am not looking at it"
Albert Einstein

The profound influence of the motion myth paradigm may have had its greatest detrimental consequence at the inception of atomic/quantum physics. Because Einstein and other physicists of his era did not expose the motion myth, atomic/quantum physics was conceived and developed within the framework of a faulty paradigm. At the time when scientists were exploring the composition of atoms and their role as the building blocks of all matter, they bypassed the role of atoms in the motion of masses through space. Atomic/quantum physics was incomplete from its inception as it built on the broad shoulders of the motion myth paradigm.

With the scientific belief that all masses are a collection of atoms, is it not logical to think that the motion of masses through space involves atoms? Until the motion myth paradigm is dissolved, modern physicists will fail to surmise a viable theory that shows how atomic/quantum processes majestically orchestrate the momentum, relativity, and gravity of masses. The answer to the question of how masses move through space in the first place remains unanswered as we move further and further into the twenty-first century.

The Missing Model of Motion

The motion myth paradigm prevented the development of a theory where the "inner" processes comprising matter balances with the "outer" forces acting upon it. This equilibrium is the absolute, fundamental key to understanding how momentum, relativity, and gravity operate from an atomic/quantum perspective. Simply put, momentum, relativity, and gravity cannot be reconciled with atomic/quantum processes without a model that explains this equilibrium process. The outcome of uncloaking the missing model of motion will be the unification of the perceived differences between Galileo's relativity, Newton's laws of motion, Einstein's relativities, and quantum mechanics.

In order to explain what this missing model of motion looks like, we first have to untangle the knot caused by the motion myth paradigm. When Newton declared that a body at rest tends to stay at rest, which is understandable from his earthbound perspective, he sustained a crucial misconception introduced by Galileo—the free ride concept. But from an atomic/quantum perspective, masses never get a free ride; all masses are always moving through space. Even objects that appear to be at rest are still moving through space— independent of the object they appear to be resting upon.

Think of astronauts experiencing a continual free fall as the space station orbits the earth. As the astronauts appear to be floating within the confines of the station, one astronaut passes an orange to another astronaut. As the orange makes the trek through space, its motion is visible by both astronauts. When the astronaut receiving the gift grabs it and places it in space next to him, the orange now appears to have stopped moving as it takes a position of rest next to the astronaut who placed it there. Is the orange really at rest? What appears to be at rest to the astronaut who just placed it in space next to him is really freefalling around the earth at 17,500 miles per hour. The orange is still in motion through space. It only appears to be resting because it is sharing the same momentum through space with the astronaut that placed the orange next to him. Objects that appear to be at rest still have "inner" driven momentum. Because Newton categorized objects at rest on the earth's surface differently than objects in motion, the free ride hinted at by Galileo became a reality for Newton and all those who stand on his shoulders.

In reality, you cannot separate the "inner" physics of bodies in motion from the "inner" physics of bodies that appear to be at rest. All objects are always in motion even when they appear to be at rest. To emphasize this point, imagine a switch that turned off the gravity that accelerates objects into the earth's surface. With the effects of gravity shut off, the slightest nudge to any of these objects would cause these objects to move away from the earth's surface. The nudge wouldn't begin the motion; it would only initiate a slight variance of the motion they already exhibited when they were moving through space attached to the hip of the earth by the effects of gravity. As these nudged objects, once bound to the earth's surface by gravity, continued in their new momentum path, moving away from the earth's surface, one would witness that these objects always had motion. Their perpetual motion was cloaked by their gravity-bound status to the earth's surface.

The atoms of masses drive momentum—even for objects that appear to be at rest. Objects can either accelerate or decelerate but at no time do they ever rest. What makes atom driven momentum so indiscernible is that bodies at rest really appear to be at rest. Gravity ties their motion to the earth's motion. This is similar to a comb sitting on a seat of a moving car sharing the same motion as the moving car. Once you slam on the breaks, it becomes apparent that the comb's motion is independent of the motion of the seat upon which it appeared to be resting. The combs momentum continues at the same rate, flying past the seat, as the seat's momentum decelerates. The motion myth misperception that objects can obtain a state of rest—a free ride—has been passed down to us and continues to veil how atomic/quantum processes are involved in the momentum, relativity, and gravity of masses.

By failing to hypothesize why masses move through space in the first place, a faulty paradigm was adopted into the scientific community, implying that masses, which are made up of atoms, just mysteriously move through space without any atomic/quantum accountability for their momentum and changes to that momentum. The motion myth paradigm continues to firmly veil an atomic/quantum explanation for the momentum, relativity, and gravity of masses. Until we uncloak an atomic model of motion, we will remain stuck in the quagmire of the motion myth paradigm.

The unsuccessful equations Einstein jotted down and scratched out with his dying hands testify that the archaic belief that masses just mysteriously move through space is grossly insufficient in a science enlightened by atomic/quantum processes. His goal of a unified field theory still remains a goal of modern physicists. Unfortunately, this goal will remain out of reach until we acknowledge the consequential impact of the motion myth paradigm.

The motion myth paradigm has had a blinding effect for centuries.

Part III

The Atomic Model of Motion
The Role of the Atom in Momentum, Relativity, and Gravity

The Illusion of Rest

When Isaac Newton declared that an object in motion tends to stay in motion, while an object at rest tends to stay at rest, he probably didn't realize he was creating a false premise that would taint the minds of physicists for centuries to come. The "illusion of rest" is the greatest stumbling block to a viable explanation of how momentum, relativity, and gravity operate from an atomic/quantum perspective.

The Atom and Motion: (*Each atom is a self-regulating confinement of energy that self-regulates changes in momentum through energy adjustments.*)

How does an object in motion stay in motion? We know that a car in motion eventually runs out of gas and stops. Why doesn't an asteroid orbiting the sun run out of gas? It just keeps going and going and going. What fuels the inertia of masses? In other words, how does an object in motion stay in motion? Do masses, such as an asteroid orbiting the sun, just mysteriously stay in motion through space without any explanation other than external forces can speed it up or slow it down? Masses get their motion from the very atoms that form their structures.

Atoms don't just mysteriously move through space. The energy making up an atom perpetually fuels its momentum through space, even when it appears to be at rest in a mass on the earth's surface. Any change in an atom's momentum is accompanied by an energy change to compensate exactly for the momentum change.

This single point, that **each atom is a self-regulating confinement of energy that self-regulates changes in momentum through energy adjustments,** forces us to rethink the role of the atom in the momentum, relativity, and gravity of masses.

26

The Atomic Model of Motion

The Atomic Model of Motion explains the role of the atom in the momentum, relativity, and gravity of masses. The atomic model of motion also sets the foundation for understanding how gravitational energy is recycled through spherical bodies.

The Atomic Model of Motion Overview

1. Quantum Momentum (or the role of the atom in the momentum of mass) is the equilibrium state of momentum where energy is neither absorbed nor emitted. This is the reason that an object in motion stays in motion.

2. Quantum Adjustments (or energy adjustments) are changes in momentum (acceleration and deceleration) that are accompanied by the absorption or emission of energy to exactly compensate for those changes. The science of motion is as exact as the science of chemistry. All changes in momentum can be explained by the addition or subtraction of energy.

3. Quantum Relativity (or how the energy of atoms and masses change as they go from one inertial frame to another) explains Galilean relativity from an atomic perspective. The sustaining energy driving all inertial frames is measurable and accountable on an atomic level using an understanding of quantum momentum and quantum adjustments.

4. Quantum Gravity (or the role of the atom in gravity): The acceleration of mass we call gravity is nothing more than the absorption of energy into the atoms of that mass. This accelerates already moving mass (atoms) in the direction of absorption. An acceleration of atoms is an acceleration of mass.

---

1. Quantum Momentum: The Role of the Atom in the Momentum of Mass

[Quantum momentum is the reason why an object in motion stays in motion. The momentum of each individual atom drives the momentum of any given mass.]

The atom is the never-ending energy that keeps all masses in never-ending motion. This is how planets continuously orbit the sun without having to be rewound or refueled. Even objects appearing to take a break on earth's surface still cruise at nineteen miles per second as gravity yields them earthbound. Turn off gravity and these objects are no longer attached to the earth's surface. Give them a little nudge and they will float off in uniform bliss until an interaction with some other form of energy redirects their route or changes their speed.

Quantum Momentum is the equilibrium state or uniform motion of any atom and is the foundation for all motion from the simplest atom to the greatest of masses. The momentum of any given atom is independent of the momentum of all other atoms. It is the bonded relationship of the independent momentum of the atoms making up a mass that gives that mass its momentum through space.

From this point forward, I will periodically use the theme of yoked horses to explain the role of the atom in momentum, relativity, and gravity.

The best way to describe the **Quantum Momentum** of an atom is to take an adventure into the nucleus of an atom. Visualize a proton and a neutron yoked together like two horses yoked together pulling a carriage. Visualize the movement of the atom as the movement of these two horses. The yoke acts like the strong force that synchronizes their independent movements. With the snap of a whip or a pull on the reigns, the horses speed up, slow down, or change directions. Now imagine that these horses never stop moving. When forces act upon them, they speed up, slow down, and or change direction, but like the horses pulling Zeus' chariot, they never run out of energy.

In the emergence of an atom, energies form the atom's structure (yoking the horses together) while simultaneously engaging the atom's momentum through space, (the synchronized movement of the two horses when yoked). The two, structure and momentum, are inseparable.

As atoms form bonds with other atoms to create masses, the momentum of each atom is linked together with the momentum of other atoms to create the momentum of the mass. The atoms that make up any mass share an order of movement that orchestrates that mass's movement through space.

A good way to visualize the **quantum momentum** of mass is to think about objects in momentum in the space station while it is orbiting the earth. Because the space station is in a free-fall, the independent momentum of all objects can clearly be observed. When an astronaut pushes an object, such as a bag of water, he is pushing all the atoms that make up that bag of water. When the force ceases, the atoms settle into the equilibrium of uniform motion. The atoms of the bag of water continue in the same direction until they bump into a wall of the station. As they run into the wall, a chain reaction transpires (which will be discussed in detail in the next section) until each atom experiences a momentum shift that leads to a new speed and direction. This new path continues until a force again acts upon all the atoms making up the bag of water. These atoms continue their new momentum uninterrupted even if the space station were to suddenly vanish from around them.

Any mass that appears to be at rest is just sharing the same motion with the object it appears to be resting on. Objects that the astronauts place in the air next to them—such as a toothbrush—stay in the same place. This is because the atoms making up the toothbrush have the same momentum through space as the atoms making up the astronaut. Their momentums are synchronized, moving at the same speed and in the same direction. Another example of the shared uniform motion of separate masses not connected to each other is objects within a car but not connected to it. As the vehicle accelerates, all the objects within the car go through internal changes to accommodate for the acceleration of the vehicle they appear to be resting in. At each new speed, all the objects in the car obtain the same uniform or quantum momentum as the car. When the car crashes into something, as its momentum is instantly changed, each object that appeared to be resting in the car continues in their own independent momentum until they crash into something such as the back of the seat, the dashboard, or the windshield. Each object has its own unique inertial momentum independent of all the other objects, including the car itself. Hence the importance of seat belts that physically attaches our momentum with the momentum of the vehicle in which we are riding.

What makes the quantum momentum of atoms and masses difficult to grasp is we continually see objects at rest on the earth's surface with no apparent motion whatsoever. We see rocks that have settled to the ground, furniture resting in houses, or a computer sitting on a desk. All of these seem to have no motion or momentum. The computer I am typing on does not appear to have any momentum so how could the atoms that make it up have momentum? This appearance of objects at rest is the same illusion that fooled Galileo and Newton and is still fooling physicists to this present day. The computer I am typing on is in a continuous state of acceleration towards the earth and only appears motionless because it is temporarily experiencing something similar to terminal velocity. Due to the effects of gravity, the uniform motion of the computer's atoms is riding on the uniform motion of the atoms of the earth. If the effects of gravity could temporarily be shut off, the momentum of the atoms of the computer through space would be more apparent. The atoms of the computer would continue to move through space at the same momentum as the atoms of the earth without being connected to the atoms of the earth because they would no longer be accelerating into them. Now if you gave the computer a push, it would slowly drift away from the earth, accentuating its own momentum through space, the synchronized momentum of the atoms from which it is made.

Quantum momentum is the inertial momentum of mass through space, which is caused by the synchronized momentum of the atoms from which mass is made. The speed and direction of mass through space remains unchanged just as Newton observed until an unbalanced force or energy acts upon the mass. Then the speed and direction are altered, but never at any time do the atoms of that mass stop or rest.

So what is the difference between an object at rest and an object in motion? Nothing. A body at rest shares the same motion as the body it appears to be resting on but their momentums through space are independent of each other. Just shut off the effects of gravity to prove this point.

2. Quantum Adjustments: Think of this section as a brief intermission that will be a useful tool to better understand quantum relativity and quantum gravity.

Quantum Adjustments are about the uneven forces that accelerate or decelerate mass. When mass accelerates or decelerates, quantum adjustments transpire within each atom to compensate exactly for these changes.

The momentum of mass does not change—a brilliant observation by Newton— unless acted upon by a force. The result of that force is a change in momentum. As changes in momentum occur, energy adjustments also transpire. Like the science of chemistry, the science of motion is exact. We just haven't figured it out yet like we have for chemistry.

As an atom's momentum increases, the atom will adjust its energy to exactly correlate with that atom's new momentum, and if an atom's momentum decreases, the atom will adjust its energy to exactly correlate with that atom's new momentum. These adjustments reestablish equilibrium within the atom as it goes through momentum changes. When the atom is in a state of equilibrium, that atom is in uniform motion and will maintain that motion until acted upon by another force or energy.

During momentum changes, each atom independently absorbs or emits the exact amount of energy to compensate for the momentum change. This regulates the momentum of an atom through space, whether the atom is by itself or is part of a collection of atoms in the form of a mass. All atoms experience quantum adjustments in energy in direct response to changes in momentum. This correlates perfectly with the conservation of mass-energy.

Quantum adjustments can better be explained by viewing each atom with a corresponding momentum pattern and energy level. The momentum pattern and energy level of an atom are two sides of the same coin that help explain changes to the continuous momentum of atoms through space. During the formation of an atom, protons, neutrons, and electrons unite in a synchronized dance to form and maintain the structure and motion of that atom through space. The structure of an atom's *momentum pattern* describes an atom's perpetual momentum through space. The *momentum pattern* of an atom remains constant unless unbalanced forces act upon it. For this reason, I refer to an atom's uniform motion through space as a *momentum pattern* because this pattern continues until it is disrupted. As an atom's momentum changes, energy is proportionately absorbed or emitted (adjusted) to compensate exactly for the momentum change. Quantum adjustments are necessary to exactly compensate for disruptions to an atom's *momentum pattern*.

A good way to describe a *momentum pattern* is to go back to the visualization of horses yoked together. Think of one proton and one neutron as a pair of horses yoked together side-by-side walking at the same speed and in the same direction. Their combined movement is their momentum pattern. If we added additional pairs of horses to the first pair, so that each pair was lined up like a team of dogs pulling a sled, the combined movement of each pair of horses would be the momentum pattern of the combined team. Notice that no matter how many sets of horses we yoke together, each individual set of horses move at the same speed and in the same direction as all the other sets of horses, but with each additional set of horses added, the overall pulling power increases. Thus, the momentum pattern describes the speed and direction of the yoked horses, notwithstanding the number of sets.

Like the yoked horses, the nucleus of an atom could be described as yoked energies that simultaneously make up an atom's structure (the yoked horses) and momentum (the speed at which they are moving) through space. This energy, the energy that makes up the protons and neutrons of the nucleus of an atom, accounts for the motion of that atom through space. Like electrons, which can absorb and emit energy, protons and neutrons absorb and emit energy that regulates their momentum through space. As the linear speed of an atom increases, energy is simultaneously absorbed into every proton and

32

neutron of that atom. As the linear speed of an atom decreases, energy is simultaneously emitted from every proton and neutron of that atom. (The electrons play a vital role in keeping the balance of energy within an atom by absorbing and emitting energy into and out of the nucleus and also into and out of the atom.)

The speed of any atom through space is determined by the *energy level* of each proton and neutron that makes up its structure or *momentum pattern*. As a force increases the speed of an atom, energy is absorbed into each proton and neutron. This increases the *energy level* of each proton and neutron of that atom. The same atom will now have an overall higher *energy level* per proton and neutron. Conversely, as a force decreases the speed of an atom, energy is emitted from each proton and neutron. This decreases the energy level of each proton and neutron. The same atom will now have an overall lower *energy level* per proton and neutron. The *energy level* of an atom (the speed the horses are moving) is in direct proportion to the energy level of each proton and neutron of an atom.

Energy level could be used in two different ways. There is the overall energy level of an atom (the total horse power) and the *energy level* of each pair of a proton and a neutron of an atom (the actual speed each individual horse is moving, which is also the same for all the horses yoked together). There is an important distinction. Let us compare an atom with one proton and neutron pair (one team of horses) to an atom with five pairs of protons and neutrons (or five teams of horses). If these two different atomic size atoms have the same momentum through space, then the *energy level* of each pair of protons and neutrons is equal (all the horses are moving at the same speed), but the overall energy level or amount of energy (total horse power) will be greater for the atom with five pairs than the atom with one pair. It will have about five times the overall energy, even though the atoms move through space at the same speed. The atom with the five pairs will have more inertia or resistance to change than the atom with one pair. Throughout the rest of this book, when I refer to the *energy level* of an atom, I am referring to the energy level of one pair of a proton and neutron. This is the *energy level* that describes its linear speed through space, notwithstanding its atomic number.

The protons and neutrons ability to absorb and emit energy allows for an atom to adjust its momentum when a force acts upon it. This holds true for each atom, notwithstanding the number of protons and neutrons in its nucleus. As the atom increases in speed or changes direction, energy is absorbed equally into every proton and neutron of the nucleus of that atom (all the horses start running faster). As the atom decreases in speed, energy is emitted equally from every proton and neutron of the nucleus of that atom (all the horses slow down). This is how the energy level (speed of one horse) and momentum pattern (the equilibrium state or steady speed of the entire team) are inseparably connected. A change in the momentum pattern of an atom (a force speeds it up or slows it down) initiates a simultaneous change in the atom's energy level (the speed at which each horse moves).

The horse example shows how the momentum pattern and energy level are actually a single process. As the *energy level* of each horse increases, the *momentum pattern* of the entire team increases by the same amount, and vice-versa. Imagine you have a team of six horses, with two horses yoked side by side, with another set yoked behind them, and the last set yoked behind them. The speed that each horse moves is the *momentum pattern* for that whole team. They all move at the same speed. When you apply a force, such as a whip to speed them up, or a pull on the reins to slow them down, you simultaneously change the energy level (speed) of each horse and the momentum pattern (speed) of the whole team. The *energy level* of one horse is the same as the *momentum pattern* for the entire team. The energy level and momentum pattern for one set of a proton and a neutron is the same energy level and momentum pattern for all the other sets of a proton and a neutron within the same atom. When you change the momentum pattern of the entire team, the energy level of each horse simultaneously changes. And vice versa, when you change the energy level of each horse, the momentum pattern of the entire team simultaneously changes. They are two sides to the same coin. This is an important concept for understanding how quantum gravity works,

Mass is a collection of synchronized atoms moving through space as the momentum pattern of each atom moves in concert with all the other atoms of that mass. Mass maintains a constant speed and direction until there is a disruption to the momentum pattern of each atom that makes up that mass. When the momentum patterns of the atoms of a mass are disrupted, energy is absorbed or emitted from the mass, from each individual proton and neutron of each atom. This reestablishes equilibrium within each atom within the mass to accommodate for the new speed and direction of that mass through space.

Visualize two masses colliding. The *momentum patterns* of the atoms of each mass are disrupted. As the *momentum patterns* are altered, the accompanying *energy levels* of the protons and neutrons of the atoms of each mass simultaneously adjust to accommodate the new *momentum patterns*. An absorption or emission of energy to exactly coincide with the degree that the *momentum patterns* were altered transpires, (each horse slows down or speeds up). As equilibrium is restored, the masses continue in their new momentums until a force or energy acts upon them again.

In summary, the momentum pattern of an atom and energy level of each proton and neutron that makes up the nucleus of an atom regulates the speed and direction of that atom through space. When momentum changes occur, quantum adjustments transpire to reestablish the equilibrium of quantum momentum. Since mass is the synchronized movement of individual bonded atoms, a momentum change in mass is a momentum change for every atom that makes up that mass. Every atom experiences a quantum adjustment. As there is no *something from nothing* in the cause and effect science of physics, changes in the momentum of a mass must be accounted for by corresponding energy changes within that same mass.

3. Quantum Relativity: How Atoms and Masses Change as They Go from One Inertial Frame to Another

Galileo learned that if a person were in a closed cabin of a ship moving at a uniform speed, he wouldn't be able to tell his relative motion to the land by appealing to the laws of physics. If he dropped an object, it would fall straight down. If he threw it up in the air, it would go straight up and straight down. In all, there was nothing he could do to determine his relative speed to land by applying what he knew about the laws of physics.

On the surface, the laws of physics appear to be the same for all inertial frames, but a deeper look into the microphysics of relativity will reveal differences on an atomic/quantum level for the same mass in different inertial frames. Since Einstein liked to use trains to explain special relativity, I will use a train example to explain quantum relativity.

If person A, traveling on a train holds a beanbag four feet above the floor and drops it to the floor of the cabin wherein he is riding, he will measure the beanbag to have fallen four feet. If person A decides to drop the same beanbag out of the moving train's window four feet above the ground, he will see the beanbag fall four feet straight to the ground in the same manner that he saw the beanbag fall four feet straight to the floor in the cabin of the train. Person B, standing on the stationary earth relative to the moving train, will measure the distance of the falling beanbag to fall more than the four feet as observed by person A. For person B, the beanbag will not only fall four feet to the ground as measured by person A, it will also fall at an angle proportional to the speed of the train. For this reason, each observer will literally measure the beanbag falling different differences. (See illustration—1). This is classic Galilean relativity, but it brings up an interesting dilemma: How can the same falling object literally travel multiple different distances such as this falling beanbag dropped and observed by person A and observed by person B? The answer to this question will be found in the *momentum pattern* and *energy level* of each atom of person A, person B, and the beanbag.

Illustration--1

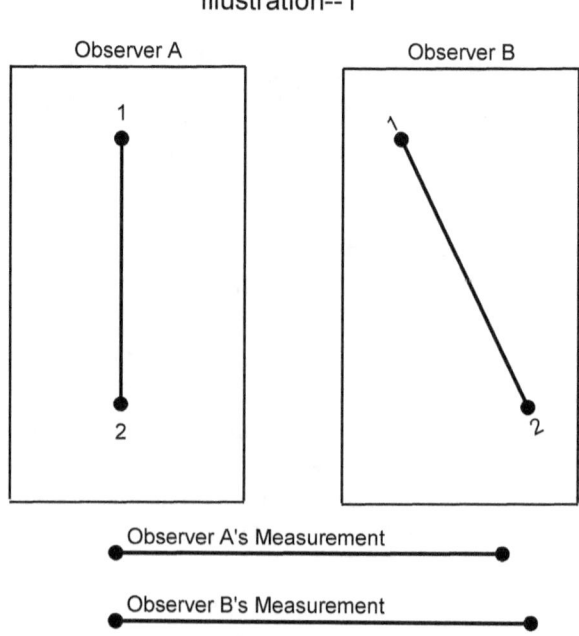

With each incremental increase of speed to the beanbag as it travels in a moving train—relative to earth's inertial frame—the more overall energy the beanbag acquires. (The hypothetical horses making up each atom run a little faster—energy is absorbed into each atom making up the beanbag.) For example, if person A drops a beanbag four feet into a pile of sand outside of a non-moving train, the beanbag will displace a certain amount of sand upon impact. The energy acquired by the falling beanbag due to the effects of gravity is transferred to the sand at impact. Now, if the train is moving and person A drops the beanbag the same four feet so it hits the same sandbox that is outside of the train, falling at an angle due to the movement of the train, the overall amount of displacement of sand caused by the falling beanbag is increased. The beanbag literally contained more energy to be transferred to the sand on impact, creating a greater displacement of sand.

This increase of energy is an actual increase of energy that is added within the very structure of every atom making up the beanbag. Any change in momentum is a change in the overall energy level of a given mass. This means that the same beanbag in different inertial frames has differing amounts of mass-energy.

How can two different observers, such as observer A and observer B, measure the same falling object to travel two different distances? The difference in distances can be explained on the atomic level. The atoms of the beanbag dropped by person A had the same momentum pattern as the atoms of the moving train and person A on the moving train that dropped the beanbag. The atoms of person B, who was standing on the earth watching the train go by, had the same momentum pattern as the atoms of the earth upon which person B was standing. The atoms of the beanbag, person A, and the train were at a higher energy level than the atoms of person B and the earth. Person A observed the beanbag from the same energy level as the beanbag, while person B observed the beanbag from a lower energy level than the beanbag. Being at the same energy level as the beanbag, person A sees the beanbag fall straight down as observed in Galilean relativity. Being at a lower energy level than the beanbag, person B sees the beanbag fall at an angle towards the ground. From their immediate energy levels in relation to the energy level of the beanbag, the measurements of both observers is absolutely correct even though the beanbag took only one absolute path through space.

When the beanbag was dropped from the moving train to the stationary earth, the atoms of the beanbag maintained the momentum pattern of the moving train until impact. Upon impact, each atom of the beanbag experienced a quantum adjustment as they switched inertial frames. In this case, the momentum pattern of each atom of the beanbag lost energy as they adjusted to become synchronized to the new momentum pattern of the atoms of the earth.

In summary, the faster a beanbag (or any mass) moves relative to the earth's inertial frame, the more energy it acquires. The energy level of each atom of a mass changes when the inertial frame that carries it changes. As a train (or any mode of transportation carrying mass) accelerates to go from one inertial frame to the next, say from 5 mph to 10 mph, its mass (and any mass that appears to be resting on it) literally acquires more energy within its atomic structures. This causes observers of differing energy levels to observe the motion of the same mass differently.

In Stephen Hawking's book, *A Brief History of Time*, Hawking states, "…suppose our Ping-Pong ball on the train bounces straight up and down, hitting the table twice on the same spot one second apart. To someone on the track, the two bounces would seem to take place about forty meters apart, because the train would have traveled that far down the track between the bounces….The positions of events and the distances between them would be different for a person on the train and one on the track and there would be no reason to prefer one person's positions to the other's" (Hawking, 1988, 17-18.)

What Hawking is admitting here is that there is no scientific explanation accounting for the differences in the distances as observed by both observers. Quantum relativity offers an explanation. Each person from their own inertial frame will observe the path of the Ping-Pong ball through space differently because the mass/atoms of each inertial frame has a different momentum pattern and energy level. The person in the train shares the same momentum pattern and energy level as the train and the Ping-Pong ball while the person on the track observes from a different momentum pattern and energy level than the train, the person on the train, and the Ping-Pong ball. Again, the cause of this phenomenon is that observers of differing energy levels will observe the motion of the same mass differently.

Why do objects that can appear to be at rest within an inertial frame follow the same laws of physics? Whether the train is at rest at the station or traveling at a uniform speed of 100 miles per hour, the Ping-Pong ball falls straight to the ground. What's going on inside the Ping-Pong ball?

38

Let's put an *atom* on the Ping-Pong table. Because the *atom* seems to be resting on the Ping-Pong table, one might assume that the imaginary horses within our *atom* are at rest, too, that when the train picks up speed, the horses can stand there and enjoy the ride. In other words, when the train goes from zero to 100, one might assume that nothing transpires within the *atom*; it is just a passive passenger along for the ride. But from an atomic model of motion perspective, when our imaginary horses within the *atom* were on the Ping-Pong table at the station, their legs were moving. The momentum pattern of the *atom* is the same as the momentum pattern of the atoms making up the Ping-Pong table, and the Ping-Pong table has the same momentum pattern as the train stationed on the tracks, and the tracks have the same momentum pattern as the earth, to which they are attached. As a force increases the speed of the train, all the atoms that are a part of the train experience changes to their momentum patterns and energy levels. Even the *atom* that appears to be resting on the Ping-Pong table experiences changes in its energy level and momentum pattern as the speed of the train increases. The imaginary horses of the *atom* resting on the Ping-Pong table change their energy level and momentum pattern to match the energy level and momentum pattern of each pair of yoked horses of all the other teams (atoms) connected to the train. All the horses of all the atoms increase their energy level. All of their feet are trotting at the same speed. (This is what is meant by there are no free rides.)

In other words, the Ping-Pong ball is not a passive passenger on the train. As the atoms within the train increase in energy, the atoms of the Ping-Pong ball will also increase in energy. Because the momentum pattern of the atoms of the Ping-Pong ball remain synchronized with the momentum pattern of the atoms of the Ping-Pong table and the rest of the train, the Ping-Pong ball will drop straight down if you held it above the table and dropped it. The effects of gravity cause the ball to fall straight down. This phenomenon happens within any inertial frame where all the atoms share the same momentum pattern (the horses are all moving at the same speed and in the same direction).

To help visualize relativity from an atomic model of motion perspective, imagine the following experiment that could be done on the international space station that is free falling around the earth. Imagine Person A floating from one side of the space station to the other side. When he reaches person B, who is stationed in the middle, he lets go of a quarter. Because the individual atoms that make up person A and the quarter share the same momentum pattern and energy level (even after person A lets go of the quarter) person A and the quarter continue to move through space at the same speed and in the same direction. Person B, who is stationed in the middle, watches as the quarter slowly moves away at the same speed that person A is moving away. Person B sees the quarter travel a greater distance from his immediate perspective than person A, who sees the quarter travel with him as if he were still holding it. Even though the quarter travels only one distinct path through space, each observer measures a different length of travel of the quarter from his or her immediate perspective. This is because Person A and the quarter remained at the same energy level, while person B observed from a different energy level.

If Galileo could have measured the energy level of the atoms of any object in his boat cabin while his boat was tied to the dock, he then could have compared that measurement to the energy level of the atoms of the same object when his boat was sailing. The measured difference would validate that as mass goes from one inertial frame to another, the momentum pattern and energy level of the atoms of that mass change to accommodate the new uniform motion. The acceleration and deceleration of masses are only temporary stages between the equilibrium of uniform motion. The acceleration or deceleration of mass is always accompanied by the absorption or emission of energy. This allows each atom to adjust to the new energy level required to maintain its new momentum pattern.

Quantum relativity is the reason the laws of physics are the same for all inertial frames.

4. Quantum Gravity: The Role of the Atom in Gravity

Gravity is typically explained as an attraction between masses. Quantum gravity explains this attraction as the emission and absorption of gravitational energy between masses. The absorbed gravitational energy accelerates the momentum of atoms. An acceleration of atoms is an acceleration of mass.

Is an object sitting on the earth's surface with no apparent motion really at rest? Newton's observation of distinguishing between objects at rest verses objects in motion also established an incurable mindset for many centuries that impeded the potential for understanding how gravity operates on the atomic level. As a result of Newton's misperception, the motion myth paradigm (the myth that mass just mysteriously moves through space) has tightly gripped the minds of most, if not all, physicists. And, to this day, physicists still cannot come to a common consensus as to how gravity operates from an atomic/quantum perspective.

When Newton said that an object in motion tends to stay in motion unless acted upon by a force, he was obviously referring to an external force such as a collision or friction. The result of the external force is a change in the momentum of that mass. From an atomic/quantum perspective, an external force causes a change in the momentum pattern of each atom, which initiates a simultaneous change in the energy level of that atom. This happens every time an external force acts upon an object in motion. Energy is absorbed or emitted to exactly correlate with the momentum change.

The same effect occurs when energy is absorbed into the nucleus of an atom. A change in the energy level of an atom initiates a change in the momentum pattern of that atom. Say the horses are spooked and they start running faster. This is a change in the *energy level* of each horse. In the same instance, the overall speed of all the yoked horses increases. This is a change in the *momentum pattern* of that team of horses. As energy is absorbed into each proton and neutron of an atom, the result is the acceleration of that atom in the direction of absorption. Its momentum simultaneously shifts to accommodate the increase in energy. This momentum shift is the acceleration.

Einstein's happiest thought was the realization that inertial acceleration and gravitational acceleration are equivalent. In other words, gravity is the acceleration of mass. He used an example of an elevator being pulled up in empty space at 32 feet per second per second to illustrate this point. (Thirty-two feet per second per second means that for every second of acceleration, an object travels an additional 32 feet of the distance of the previous second. At the end of the first second, an object travels 32 feet. For the second second, the object travels 64 feet within that second for a total of 96 feet for the first two seconds (32+64). For the third second, the object travels 96 feet within that second for a total of 192 feet (96+96) and so on.) The elevator being pulled up in empty space would create the same gravitational acceleration that a person experiences standing on the earth or an object falling towards the earth. The reality that gravity is the acceleration of mass creates the basis for quantum gravity.

From an atomic perspective, inertial acceleration is equivalent to gravitational acceleration. Both result in the change of the energy level and momentum pattern for every atom involved. An external force causes inertial acceleration, whereas the absorption of energy causes gravitational acceleration. Whether by an external force or the absorption of energy, the result is the same—the atom or mass accelerates. This is because a change in the momentum pattern (external force) changes the energy level of an atom, and vise versa, a change in the energy level (absorption of energy) changes the momentum pattern of an atom. Either way, the result is acceleration. The simultaneous effect of the momentum pattern changing the energy level or the energy level changing the momentum pattern is like blowing air into a balloon. As the balloon receives more air, the boundaries of the balloon expand. As the balloon loses air, its boundaries contract. As pointed out in the quantum adjustment discussion, an external force can accelerate or decelerate the linear speed of an atom, but energy absorption always accelerates the linear speed of an atom.

Since mass is a collection of bonded atoms, an acceleration of atoms is an acceleration of mass. The constant flow of energy into the atoms of a mass creates gravitational acceleration. The absorbed energy changes the energy level of each atom, which simultaneously changes the momentum pattern of each atom. This accelerates the already moving mass in the direction of absorption.

This is why Newton's misperception was so crucial. If mass is perceived to be at rest when it is sitting on the earth's surface, then the atoms making up that mass are not perceived to be the cause of that mass's motion through space. Newton's misperception assumed that mass is getting a free ride on the earth. This misperception blocks the potential for understanding how gravity really works. By correcting Newton's misperception, we can understand that the atoms that make up any mass are the cause and continuation of that mass's motion through space, (the horses are always moving). What appears to be mass resting on the earth's surface is actually mass in motion wanting to accelerate but experiencing something akin to terminal velocity due to the resistance of the earth's surface. If we could shut off the effects of gravity—the energy causing acceleration—the mass would no longer be accelerating into the earth, nor would it be attached to the earth. A good nudge would cause the mass to go off in a different direction than the earth, accentuating the motion it already had.

The effects of absorbed energy causing gravitational acceleration can be seen when an accelerating object decelerates at the moment of impact with a lower energy level surface such as the earth's surface. In the process of reestablishing equilibrium with atoms of a lower energy level, quantum processes transpire. This sets off a chain of events that can be witnessed at the moment of impact as will be explained in the falling penny example.

The Falling Penny

An example of the effects of quantum gravity is dropping a penny to the floor. First of all, as you hold the penny above the floor, the penny is already in motion. It is sharing the same motion as the hand holding it, just as the quarter shared the same motion as the moving astronaut, even after he let it go. When you let go of the penny, the atoms making up the penny absorb energy emanating from the surface of the earth. (The energy emanating from the earth's surface will be addressed later.) The energy level of each atom changes—like the snap of a whip causing each horse to run faster. This simultaneously changes the momentum pattern of each atom, (the whole team is moving at a faster pace). This increases the overall motion of the penny in the direction of absorption 32 feet per second per second until it hits the ground.

When the penny hits the ground, its increased energy level from the absorbed energy that accelerated its momentum interacts with the lower energy level of the ground. At the very moment of impact, the momentum pattern of each atom of the penny changes. Energy is emitted from each atom to begin the accommodation of each atom's towards the new momentum pattern of the atoms of the earth. At the initial bounce, the penny is off in a different direction. If not for gravity, the penny would appear to float off in its new direction until a force acted upon it. (To help visualize this, think of an object in the space station that is floating horizontally towards one of its walls. When the object hits the wall, it changes its direction, and then continues in its new direction until a force acts upon it.) As the penny starts to go off in a different direction, it begins to absorb more energy emanating from the earth's surface, changing its direction and accelerating it towards the earth again to repeat the process over and over until the penny finally appears to rest on the earth's surface, with its atoms sharing the same momentum pattern as the atoms making up the earth. During this whole process, the sum of mass-energy in the universe remains unchanged.

When the penny makes contact with the lower energy level surface, there is an immediate change in the energy level of every atom of the penny. The atoms of the surface where the penny landed are also disrupted at impact, causing the absorption and emission of energy according to each atom's equilibrium need. (This is like the sand that was displaced in the sandbox when the beanbag landed in it.) When the penny finally appears to rest on the earth's surface, the momentum pattern and energy level of the atoms of the penny remain synchronized with the momentum pattern and energy level of the atoms of the earth until another disruption occurs such as when somebody picks up the penny.

Why does the penny appear to stay at rest on the earth's surface? Even though it is continuously absorbing energy emanating from the earth's surface, causing it to want to accelerate, its continuous contact with the surface of the earth restricts the acceleration. In other words, it has reached terminal velocity. The energy necessary for acceleration is continuously being absorbed into the penny, but because the momentum pattern of the atoms cannot change due to their terminal velocity, the absorbed energy is emitted at the same rate of absorption.

Drop a penny a few times on the surface of a table. Watch at impact as the penny goes from a higher energy level (the energy acquired to accelerate it towards the surface of the table) to the lower energy level of the surface of the table (the previous energy level of the penny before you picked it up and dropped it). Drop the penny from various heights to compare the amount of energy it acquires from the varying heights. As the penny hits the table and bounces a few times and then vibrates until it eventually comes to rest on the table, you are witnessing the atoms of the penny going from a higher momentum pattern and energy level to the lower momentum pattern and energy level of the atoms of the table.

Gravity is not a mysterious force of attraction between masses, nor is it the effects of warped space-time, but rather, it is energy (forces) acting within matter. It is the effect of absorbed energy on already moving mass. This is why exposing Newton's misperception that objects can be at rest is so important to understanding how gravity really works. Without this step, you cannot understand how gravity is the absorption of energy accelerating mass that already has momentum.

This is why Einstein was unable to unify general relativity (his explanation of gravity) with atomic/quantum processes. Like Newton, he failed to recognize that all objects are always in motion, the first major step for understanding how gravity operates from an atomic/quantum perspective. Instead, he brilliantly created space-time as a substitute for atomic/quantum processes, accurately predicting the effects of gravity without addressing how atomic/quantum processes are involved in the intricate process of gravitational acceleration.

Einstein's space-time describes distortions in space caused by large masses, such as suns and planets, and the effect this has on moving objects near their surfaces, such as other planets, moons, asteroids, and light. In reality, these distortions are actually higher concentrations of energy emanating from the surfaces of large spherical masses such as the sun or the earth. (This will be explained in more detail later.)

Quantum gravity can now be easily explained as the acceleration of mass caused by the absorption of gravitational energy. The absorbed energy changes the energy level of each atom, which simultaneously changes the momentum pattern of each atom, causing acceleration towards the source of the energy being absorbed.

Because the momentum of each proton and neutron is independent of the momentum of all other protons and neutrons (like each set of yoked horses connected together with other yoked horses) the number of protons and neutrons in an atom doesn't change that atom's acceleration rate in comparison to atoms of differing numbers of protons and neutrons, (atomic number). All protons and neutrons of a mass are exposed to the same energy source and absorb the same amount of energy, causing the same momentum shift in the direction of absorption. (All the horses increase their speed at the same rate.) This is why Galileo could drop two differing size balls off the same tower and watch them land at the same time. Because the larger ball has more protons and neutrons, its overall energy or inertia will be greater than the ball with fewer protons and neutrons. But the acceleration of both balls will be the same because all protons and neutrons accelerate at the same rate, for all protons and neutrons are exposed to the same energy concentration emanating from the earth. For this reason, all atoms, elements, compounds, and masses accelerate at the same rate within the same gravitational field.

The Visible Effects of Quantum Gravity

The absorbed energy due to the effects of gravity is visible. When I hold a pencil several feet above a hard tile floor and drop it, the absorbed energy responsible for accelerating the atoms of the pencil towards the floor causes the pencil to bounce a few times when it hits the floor. Energy is emitted from the atoms of the pencil with each bounce until the atoms of the pencil share the same energy level as the atoms of the surface of the floor upon which it now appears to be resting. When I held the pencil above the floor, it is said to have potential energy. Potential energy is nothing more than potential changes in momentum patterns—acceleration—before a terminal velocity is reestablished with the surface of the earth—deceleration.

When a ball bounces off the ground, it moves away from the surface of the earth. It would continue in this direction uninterrupted if it weren't for energy being absorbed back into the atoms of the ball due to the effects of gravity. After the ball reaches its apex, briefly sharing the same momentum pattern and energy level as the atoms of the earth, it begins to accelerate towards the earth again. With each bounce, more energy is emitted until eventually the ball comes to a resting position on the earth. You can visually see the effects of acquired energy causing the acceleration that precedes each bounce. When the ball appears to be resting on the earth's surface, it is still moving through space. The momentum pattern and energy level of each atom of the ball are still the source and cause of its movement through space, but now they are synched with the momentum pattern and energy level of the atoms making up the surface of the earth. The ball now appears to our eyes to be talking a position of rest on the surface of the earth. If I shut off the energy emanating from the earth and then nudged the ball towards the direction of the sky, it would move away from the earth like a helium balloon escaping from the grasp of a child, accentuating the independent momentum of the atoms of the ball from the atoms of the surface of the earth.

Absorbed energy accelerates mass towards the surface of the earth and emitted energy accompanies the deceleration of mass as it eventually nestles into what appears to be a resting state on the surface of the earth.

The Imaginary Wall in Space

An example to help visualize quantum gravity is to imagine a wall in space emitting gravitational energy. If you placed a marble several feet away from the wall, it would begin accelerating towards the wall. When it reached the wall, it might bounce a few times and then the marble would appear to rest against the wall. The energy level and momentum pattern of each atom of the marble would be at a terminal velocity against the wall, wanting to accelerate but restricted by the wall. The atoms of the marble would share the same momentum through space as the atoms that make up the wall. In this state, the same applied force would move the marble the same distance in any direction on the wall. (In essence, the wall acts like the surface of the earth.) If you flicked the marble with the same force in any direction, it would go the same distance. If you pulled it a few feet away from the wall and let it go again, it would again start accelerating towards the wall. When it hit the wall, it would bounce a few times, emitting energy until the energy level and momentum pattern of the atoms of the marble matched the energy level and momentum pattern of the atoms of the wall. Then it would again be in a state of terminal velocity against the surface of the wall, appearing to be at rest on the wall.

Tides

Another effect of gravity that is observable but difficult to explain is the cause of tides. Heretofore explained as the gravitational pull of the moon on the earth can now be explained as concentrated gravitational energy emitted from the spherical shaped moon and absorbed by the atoms making up the water. This causes a shift in their momentum patterns, accelerating them in the direction from which the energy was absorbed until a terminal velocity is reached between the energy emitted by the moon in contrast to the energy emitted by the earth. The moon doesn't mysteriously attract the water nor does warped space cause the water to move towards the moon, but rather, it is energy emanating from the large, spherical-shaped moon that accelerates the atoms of the water towards the direction of absorption.

Orbits

Whether it is our sun orbiting the center of our galaxy, the earth orbiting our sun, or the moon or satellites orbiting the earth, they all seem to follow the same simple pattern. Quantum momentum keeps an object in a perpetual state of motion. Quantum gravity keeps that same object in a perpetual state of acceleration, keeping it in an orbital pattern around a spherical body. It is precisely because these objects already have momentum that energy emanating from the surfaces of large spherical bodies can accelerate this momentum into an orbital pattern.

47

Summary

Gravity, previously described as an attraction or later as warped-space, can now be described as the absorption of gravitational energy into the atoms of a mass, initiating and sustaining an acceleration of already moving mass in the direction from which the gravitational energy is being absorbed.

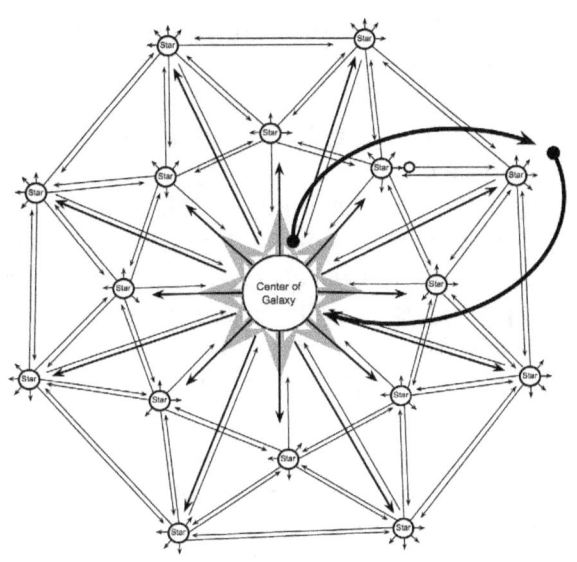

Part IV
The Galaxy Gravity Cycle (GGC)
The Role of Spherical Bodies in Absorbing and Redirecting the Flow of Gravitational Energy within a Galaxy

How does the earth (or any spherical body) sustain the constant effects of gravity?

Thanks to the effects of quantum momentum, the earth, as a spherical body, steadily stays in motion through space. The amount of gravitational energy it receives from the sun allows it to continuously accelerate towards the sun in an orbital pattern. What happens to the gravitational energy absorbed into the earth? How does this energy create earth's gravitational effect? How does emitted energy from spherical bodies better explain the warped space of general relativity? And how does the Galaxy Gravity Cycle correct the erroneous foundation of modern physics?

What happens to the gravitational energy absorbed into the earth?

The gravitational energy absorbed into the earth that keeps the earth accelerating into the sun is eventually emitted back through the spherical

49

surface of the earth, creating the effects of gravity that we experience living on the earth. How does this work?

The core and layers of the earth, or of any spherical body, is the catalyst that maintains the cycle of gravitational energy flowing into and out of spherical bodies. As mentioned in quantum gravity, gravitational energy absorbed by an atom causes it to accelerate towards the direction of absorption. As atoms of a spherical body absorb gravitational energy, they are restricted in the amount of acceleration they can experience. This is due to their proximity with the other atoms within the body. The gravitational energy not integrated into one of these confined atoms of the spherical body due to their inability to accelerate is simultaneously emitted from that atom. The emitted gravitational energy can then be absorbed into a nearby atom. But as the nearby atom's acceleration is also limited, it simultaneously emits the gravitational energy not being integrated into it. This process repeats itself over and over again within the body of compactly confined atoms. Eventually, the gravitational energy that has nowhere to be absorbed within the spherical body is emitted through the surface area of the spherical body and out into space.

The gravitational energy emanating from the surface area of the spherical body creates the gravitational effects that spherical bodies exhibit near their surfaces. As the gravitational energy being emitted through the spherical surface goes out into space, it causes other masses to accelerate towards the spherical body that is emitting the gravitational energy. It is important to note that the concentration of gravitational energy will be greater near the surface of the spherical mass and will defuse as it goes further away from the spherical surface, (See Illustration—2 Energy Emitted from Spherical Surface Area). In Summary, as energy is absorbed into a spherical body, the atoms either accelerate or—like the penny experiencing a process similar to terminal velocity as it rests on the earth's surface—acceleration is restricted and energy is emitted at the same rate of absorption. The unabsorbed gravitational energy eventually makes it back to the spherical surface and is emitted out into space. This constant flow of emitted gravitational energy accelerates masses towards the spherical surface and keeps atoms, molecules, and masses earthbound.

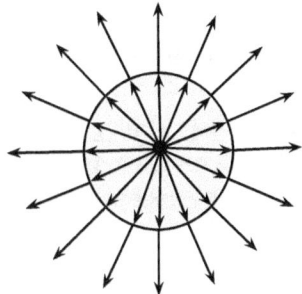

Illustration--2

Energy Emitted from Spherical Surface Area

Due to a state comparable to terminal velocity, restricting the acceleration of atoms within a spherical body, absorbed energy is simultaneously emitted from the atoms comprising the spherical body until the energy finds its way to the surface of the spherical body and is emitted out into space. This creates concentrated amounts of energy near the surface of a spherical body. As this energy moves farther away from the surface, its concentrated amounts are proportionately diluted as illustrated by the distance of the arrows above getting farther and farther apart the farther they move away from the origin or center of the spherical body. Thus, the gravitational field is strongest near the surface of the spherical body and gets proportionally weaker the farther out in space you go from the spherical surface.

Why do the atoms near the surface of the earth stay earthbound? Shouldn't they just independently accelerate towards the sun? As the atoms of the earth facing the sun receive gravitational energy from the sun, they accelerate towards the sun. This is most evident with the fluidity of water in the case of tides. As water molecules receive gravitational energy from the sun and moon, they begin to accelerate towards the sun and moon. At the same time, the gravitational energy recycled through the spherical body of the earth and emitted through the surface area of the earth accelerates the atoms on the surface of the earth towards the surface of the earth, keeping them earthbound. This simultaneous tug-a-war creates the phenomenon of tides and explains why tides are at their highest when the sun and moon are aligned but why the ocean waters and all other atoms, molecules, and compounds always stay earthbound.

How do spherical bodies increase in their size and mass? The gravitational energy emitted from a spherical body causes other masses to accelerate into the spherical body. This adds more mass to the spherical body, increasing the circumference of the amount of atoms that are restricted in movement, increasing the spherical size and surface area emitting gravitational energy. This creates a greater concentration of gravitational energy near the growing surface area and a stronger gravitational field surrounding the spherical body. As a spherical body increase in size, it simultaneously increases the concentration of gravitational energy being emitted from its surface area.

How does this energy create earth's gravitational effect?

The cycle of gravitational energy being absorbed into and eventually emitted out of a spherical body creates that spherical body's gravitational effect. As stated above, the space necessary for atoms to change their momentum within a spherical body is limited. As a constant flow of gravitational energy accelerates atoms within a spherical body, the amount of acceleration they can experience is limited due to the compact nature of all the atoms making up the spherical body. Each atom of the spherical body then experiences a process similar to terminal velocity. At terminal velocity, the atoms absorb and emit energy at the same rate. Eventually, the gravitational energy emitted by the limited accelerations of the atoms that make up the spherical body has nowhere to escape except through the surface of the spherical body. When it reaches the surface, it is then emitted into space. This causes any object on or near the surface of the spherical surface to accelerate towards the surface. This cyclical process of spherical bodies like the earth absorbing and emitting gravitational energy keeps a constant flow of gravitational energy moving into and out of a spherical body. And the energy continuously being emitted through the spherical surface into space creates the gravitational effect associated with the spherical body. The denser the concentration of this energy being emitted, due to a greater size and mass of the spherical body, the greater the gravitational effect in accelerating atoms, molecules, and masses on or near the surface. And as gravitational energy emitted through the surface of a spherical body continues to go further out into space, so does the gravitational effect of that energy. The farther away the gravitational energy moves away from the spherical body, the weaker the gravitational field surrounding that spherical body. But as long as any of this gravitational energy is absorbed into the atoms of masses it penetrates, those atoms of masses will accelerate towards the spherical surface from whence this energy was emitted. Again, Illustration—2 shows how the gravitational energy emitted from a spherical body is diluted as the gravitational energy spreads out the farther it gets away from the spherical surface.

How does emitted energy from spherical bodies better explain the warped space of general relativity?

The size and mass of a spherical body determines the amount of energy flowing into it. This in turn determines the concentration of energy exiting its spherical surface. The energy exiting a spherical surface is the warped space-time of general relativity. The gravity effect is greatest around the surface area of spherical bodies and weakens the further you move away from the surface. (See Illustration—2 Energy Emitted from Spherical Surface Area.) What Einstein mathematically composed as warped space-time in his general relativity theory is actually the effects of concentrated gravitational energy exiting through a spherical surface. As this energy moves away from the spherical surface, its concentration is diluted more and more the farther it moves away from the spherical surface, but when any of this energy is absorbed into the atoms of a mass, it will cause those atoms to accelerate. Einstein's general relativity theory is a mathematical description of higher and lower concentrations of gravitational energy exiting through spherical surfaces.

Electromagnetic waves are not impervious to Gravitational Energy. As a photon passes through a field of high concentration of gravitational energy being emitted through a spherical surface, gravitational energy is absorbed directly into the photon. Instead of increasing acceleration like it does when gravitational energy is absorbed into the nucleus of an atom, the photon experiences a frequency shift. The photon now has more energy in the form of a higher frequency. The speed of the photon didn't change, but the frequency shift caused a slight altering of its path in the direction from which the energy was absorbed. The bent light around spherical bodies is the result of successive frequency shifts within the photons of that light as the light passes through a higher concentrated gravitational energy field.

A proof for the validity of the Atomic Model of Motion (AMM) and the Galaxy Gravity Cycle (GGC) hinges upon the starlight displaced by the gravity of the sun. Einstein pointed to the warping of space-time as the cause of this validated phenomenon, (Eddington, 1919). The bending of light around large massive objects like the sun is not caused by warped space, but rather, it is caused by large concentrations of gravitational energy emanating from the surfaces of spherical bodies like the sun. As light passes by the surface of the sun, each photon absorbs some of this gravitational energy, causing a frequency shift. Without affecting its speed, this frequency shift slightly changes its direction towards the source that initiated the shift. The displaced starlight will be blue shifted from the same starlight that isn't displaced, validating the underlying principle from which the Atomic Model of Motion (AMM) is built upon: Energy adjustments, such as the absorption or emission of some from of energy, must accompany every speed and or directional change of every atom or photon, otherwise, the conservation of mass-energy would be violated.

How Does the Galaxy Gravity Cycle Correct the Erroneous Foundation of Modern Physics?

It may be a hard pill to swallow to think that modern physics is built upon an erroneous foundation. In the emergence of atomic/quantum physics, there is a clear and distinct break from classical physics. In this split, a new science emerges that is distinct and separate from the old science. Yet, the new science cannot account for all the actions and reactions that take place in the old science, mainly momentum, relativity, and gravity. Thus, in physics, we are left with a science for the macro, (general relativity), and a difference science for the micro, (atomic/quantum processes). The Galaxy Gravity Cycle (GGC) and the Atomic Model of Motion (AMM) bring these two separate sciences together. The Galaxy Gravity Cycle replaces the warped-space of General Relativity while the Atomic Model of Motion introduces the role of atoms in the momentum, relativity, and gravity of masses. Together, they bridge the quantum gap by introducing how atomic/quantum processes are involved in the momentum, relativity, and gravity of masses and how these processes operate within the cyclic flow of gravitational energy within a galaxy. The Galaxy Gravity Cycle and the Atomic Model Motion take us closer to what Einstein was seeking:

> "For decades Einstein attempted to develop a unified field theory...connecting the movement of planets and stars with the operations of the tiniest subatomic particles." (Lacayo, 2014, 9)

The motion myth is the erroneous foundation upon which modern physics is built. That is why atomic/quantum physics had to break off of classical physics in order to progress in its development. At the conception of atomic/quantum physics, there were no "inner" hypothesizes to explain the potential role of the atom in the momentum, relativity, and gravity of masses. By exposing the motion myth and introducing two models that work in tandem, it is my hope that the science of motion—momentum, relativity, and gravity fused with atomic/quantum processes—can be further developed into a single, unified science that succeeds in accomplishing what Einstein started as stated in the above quote.

The science of motion must be as exact as the science of chemistry.

In the end, the purpose of science is not to compose reality to match our perceptions of it, but rather to change our perceptions until they conform to reality—to see things as they are even if it is different from how we want them to be.

Appendix A: The Problem with Special Relativity

Does a Photon Have the Same Relativistic Qualities as an Atom?

Quantum relativity as explained in this book as part of the Atomic Model of Motion (AMM) would be incomplete without explaining the problem with Einstein's Special Relativity. Einstein's dilemma was similar to the beanbag scenario. Replace the beanbag with a light pulse. Person A, on the moving train, sees a light pulse go four feet. Person B, standing on the stationary earth relative to the moving train, sees the same light pulse travel more than the four feet. (Please note that I understand the above example is not realistic and could not be observed with the natural eye. The scenario is based on the premise that if the distance the light pulse traveled was a lot longer, both observers would measure the same light pulse traveling a different distance.) This is because the light pulse will not only travel the four feet to the same target, but it will also travel at an angle proportional to the speed of the train, just like the beanbag. Steven Hawking put it like this:

> Since the speed of the light is just the distance it has traveled divided by the time it has taken, different observers would measure different speeds for the light. In relativity, on the other hand, all observers *must* agree on how fast light travels. They still, however, do not agree on the distance the light has traveled, so they must therefore now also disagree over the time it has taken. (The time taken is the distance the light has traveled – which the observers do not agree on – divided by the light's speed – which they do agree on.) In other words, the theory of relativity put an end to the idea of absolute time! (Hawking, 1996, 21-22.)

This scenario only exists if you assume light has the same relativistic qualities as mass (atoms), meaning that mass (atoms) can obtain an apparent state of rest in any inertial frame. Without understanding the role of the atom in quantum relativity, one could easily make this mistake.

56

To answer the dilemma proposed by Stephen Hawking, I ask the following question: *What is the major difference between a beanbag (a collection of atoms) and a light pulse?* A beanbag (atoms) can appear to be at rest in any inertial frame. On the other hand, an emitted light pulse is never at rest in any inertial frame. This means that an emitted light pulse follows a different set of rules in how it moves through space than atoms in the form of a beanbag. For example, it is common knowledge that the speed of light is unaffected by the speed of the mass (atoms) emitting it, but the speed of an object (atoms) is directly affected by the speed of the vehicle (more atoms) from which it is released. This major difference is due to the fact that in Galilean and Newtonian physics, objects (atoms) can acquire a perceived state of rest within an inertial frame—meaning synchronized momentum patterns—whereas a freely moving light pulse never finds rest within any inertial frame.

Let's quickly diagram the problem.

Illustration--3

Light Clocks

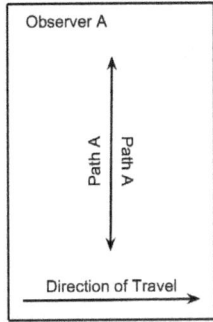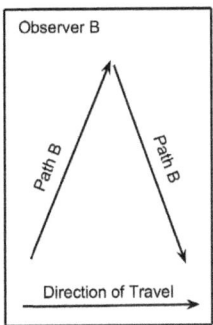

If *observer A* is traveling in a train with a hypothetical light clock, the light will move in an up and down motion in reference to *observer A* (Path A). *Observer B*, who is stationary to the motion of the train that is carrying *observer A* and the light clock, will observe the path of the light moving at angles in the direction of the uniform motion of *observer A* and the light clock (Path B). The distance of *path A* is different than the distance of *Path B*. Yet, from each person's perspective, they are both correct. This is quite a paradox if the speed of light is constant for all observers and that the same light pulse can also be measured to travel two or more different distances. Einstein's idea was to take the different distances of travel and divide them by the same light speed; you end up with two different times for the same event. Mathematically, it makes sense. Illustration—3 is a classic example used to explain time dilation of Special Relativity in many up-to-date encyclopedias. In illustration—3, if you replaced the light pulse with a bouncing ball, this would be an example of classic Galilean relativity for which I have already provided an explanation on a quantum level. However, a light pulse and a bouncing ball are not the same, so you shouldn't expect the same results. Einstein's mistake was to assume that a light pulse operated by the same rules as atoms in Galilean relativity.

Einstein failed to conclude that if the speed of light is independent of the motion of the mass emitting it (Alväger et al. 1964) then its direction or path should also be independent of that motion.

The **speed** and **direction** of emitted light pulses are unaffected by the motion or momentum of the objects emitting them. An object of mass, on the other hand, which can be at the same momentum level as the observed inertial frame within which it is contained, is directly affected by the speed of the object releasing it. Atoms, which make up objects and masses, must be treated differently than freely moving light pulses, which are never at the same momentum level of any observed inertial frame. In other words, freely moving light pulses cannot be regarded in the same manner as masses (atoms) in respect to Galilean relativity and inertial frames.

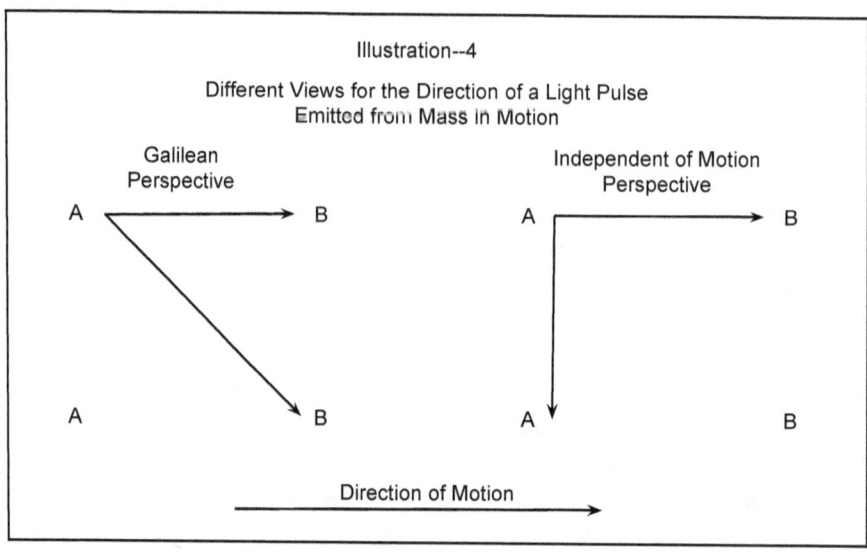

Illustration--4

Different Views for the Direction of a Light Pulse
Emitted from Mass in Motion

In illustration—4, *A* represents uniformly moving mass in the direction of *B*; *A* also represents the emission of a light pulse perpendicular to the motion of the mass emitting it. The *Galilean Perspective* demonstrates light's direction or path of travel *dependent* on the motion of the mass emitting it—a continuation of Galilean relativity. The *Independent of Motion Perspective* demonstrates light's direction or path of travel independent of the motion of the mass emitting it.

Just as the **speed** of light is unaffected by the motion of the mass emitting it, the **path or direction** it travels after its emission should also be unaffected by the motion of the mass emitting it. This is because light does not have relativistic qualities, whether confined within mass or freely flowing in space. It goes back to comparing the motion of a photon to the motion of an atom. A photon has one fixed speed through space as calculated by Maxwell, whereas an atom's speed through space varies according to the momentum pattern and energy level of that atom. For this reason, when mass is pushed away from a moving body, its new momentum is a continuation of its previous momentum, whether dropping a package out of a flying airplane or throwing a baseball from a moving vehicle. On the contrary, the speed of light is unaffected by the speed of the object emitting it, such as turning a flashlight on from the same moving vehicle from which you threw the baseball.

I remember someone telling me how Special Relativity was demonstrated in a college class. One student took a piece of chalk and perpetually drew a line going up and down while staying in one place. The second person did the same up and down motions as the first person while walking along the chalkboard from one end to the other end. This supposedly demonstrated how light travels different distances in different inertial frames, creating the need for the contraction of time to account for the differences in distances. Unfortunately, this isn't how light works. A photon does not have the same relativistic quantities as an atom. This means that the speed of light is not the same for differing inertial frames (all moving bodies) as postulated by Einstein (who must have assumed that light had the same relativistic qualities as an atom), but rather, the speed of light operates independently of all inertial frames. Once light is emitted from the confined energies of inertial mass, its *speed* and *direction* of travel move independently from the momentum of the mass emitting it.

Of all people, why would Einstein make this crucial mistake and postulate that the speed of light is the same for all inertial frames? He was stuck in a motion myth paradigm, the idea that objects just mysteriously move through space. He failed to theorize the role of the atom in momentum, relativity, and gravity. For Einstein, because objects just mysteriously move through space, Galilean relativity remained a mystery. No one, including Einstein, could scientifically explain why the laws of physics are the same for all inertial frames, (quantum relativity). In his ignorance, Einstein assumed that if the laws of physics are the same for all inertial frames, then the fixed, unchanging speed of light as calculated by Maxwell must also be the same for all inertial frames. It was a terrible miscalculation on Einstein's part that reflected the scourge of the motion myth paradigm.

When Galilean relativity is explained from a quantum perspective using momentum patterns and energy levels, then one can easily see that a photon operates differently than an atom. An atom can obtain an apparent state of rest in differing inertial frames, whereas a photon never obtains an apparent state of rest in any inertial frame. Had Galileo been able to observe the energy levels of atoms, he would have noticed that they change to correspond with differing inertial frames, (AMM).

But alas, in 1905, Einstein's application of time (with the help of the Lorentz transformation) in his paper *On the Electrodynamics of Moving Bodies* delayed the discovery of the role of the atom in the momentum, relativity, and gravity of masses.

Bibliography

Bar, Robert. *Stephen Hawking, best-known physicist of his time, has died.* Associated Press. March 13-14, 2018.

Brian, Denis. *Einstein: A Life.* New York: John Wiley & Sons, Inc. 1996.

Hawking, Stephen W. *A Brief History of Time.* New York: Bantam Books. 1988.

Hawking, Stephen W. *A Brief History of Time.* New York: Bantam Books. 1996.

Lacayo, Richard. Time Home Entertainment. *Albert Einstein: The Enduring Legacy of a Modern Genius.* New York: Time Books. 2014.

Rovelli, Carlo. Seven Brief Lessons on Physics. Penguin Books. 2014.

Seifer, Marc J. *Wizard: The Life and Times of Nikola Tesla: Biography of a Genius.* New Jersey: Carol Publishing Group. 1996.